TOUGH TRUCKS

RAM™ TRUCKS

BY SETH LYNCH

Gareth Stevens
PUBLISHING

HOT TOPICS

Please visit our website, www.garethstevens.com. For a free color catalog of all our high-quality books, call toll free 1-800-542-2595 or fax 1-877-542-2596.

Cataloging-in-Publication Data
Names: Lynch, Seth.
Title: Ram trucks / Seth Lynch.
Description: New York : Gareth Stevens Publishing, 2019. | Series: Tough trucks | Includes glossary and index.
Identifiers: ISBN 9781538231500 (pbk.) | ISBN 9781538230367 (library bound) | ISBN 9781538231517 (6 pack)
Subjects: LCSH: Trucks--Juvenile literature.
Classification: LCC TL230.15 L93 2019 | DDC 629.224--dc23

First Edition

Published in 2019 by
Gareth Stevens Publishing
111 East 14th Street, Suite 349
New York, NY 10003

Designer: Katelyn E. Reynolds
Editor: Kristen Nelson

Photo credits: Cover, p. © iStockphoto.com/duckycards; cover, pp. 1–32 (background) phiseksit/ Shutterstock.com; p. 5 Dong liu/Shutterstock.com; pp. 7, 30 (1920s) Buyenlarge/Getty Images; p. 9 Arthur Siegel/Anthony Potter Collection/Getty Images; pp. 11, 30 (1950s) Riley (https:// www.flickr.com/people/64583732@N04) from Christchurch, New Zealand/NZ Car Freak/ OSX II/Wikipedia.org; pp. 13, 30 (1960s) AutoPhoto/Wikipedia.org; p. 14 Jim R. Bounds/ Bloomberg via Getty Images; pp. 15, 30 (1980s) G.Goodwin Jr. & Snark/Slant6guy at English Wikipedia/Captain-tucker/Wikipedia.org; pp. 17, 30 (1990s) Car Culture, Inc./Car Culture ® Collection/Getty Images; pp. 19, 30 (2000s) Art Konovalov/Shutterstock.com; p. 21 Bill Pugliano/ Getty Images; pp. 23, 30 (2019) Scott Olson/Getty Images; p. 25 Steve Lagreca/Shutterstock.com; p. 27 Ed Aldridge/Shutterstock.com; p. 29 Dana60Cummins/Wikipedia.org.

Printed in the United States of America

CPSIA compliance information: Batch #CW19GS: For further information contact Gareth Stevens, New York, New York at 1-800-542-2595.

CONTENTS

IT STARTS WITH DODGE

Ram trucks can trace their history back more than 100 years to the Dodge Brothers Company. It began making **vehicles** in the early 1900s. Today, Ram is its own **brand** and a maker of some of the best **pickup trucks** on the road.

Ram 1500

FULL THROTTLE!

When the company began
in the early 1900s,
Dodge first made stove parts.
Then, they began making car parts.

Dodge put out its first trucks in 1918, though they were mostly for **commercial** use. Soon, Dodge was selling a 1 1/2-ton truck and then a 3/4-ton truck starting in 1924. Car company Chrysler bought Dodge in 1928. Dodge continued making commercial trucks into the 1930s.

FULL THROTTLE!

Dodge trucks were helped by being part of Chrysler. They used well-made, powerful engines, or the machines used to power cars, from other Chrysler vehicles.

SERIES OF TRUCKS

During World War II (1939–1945), Dodge made many different trucks for American forces. The B-Series trucks, introduced after the war in 1948, made Dodge's name stand out in the truck market. The modern look and performance of the B-Series trucks made them sell well.

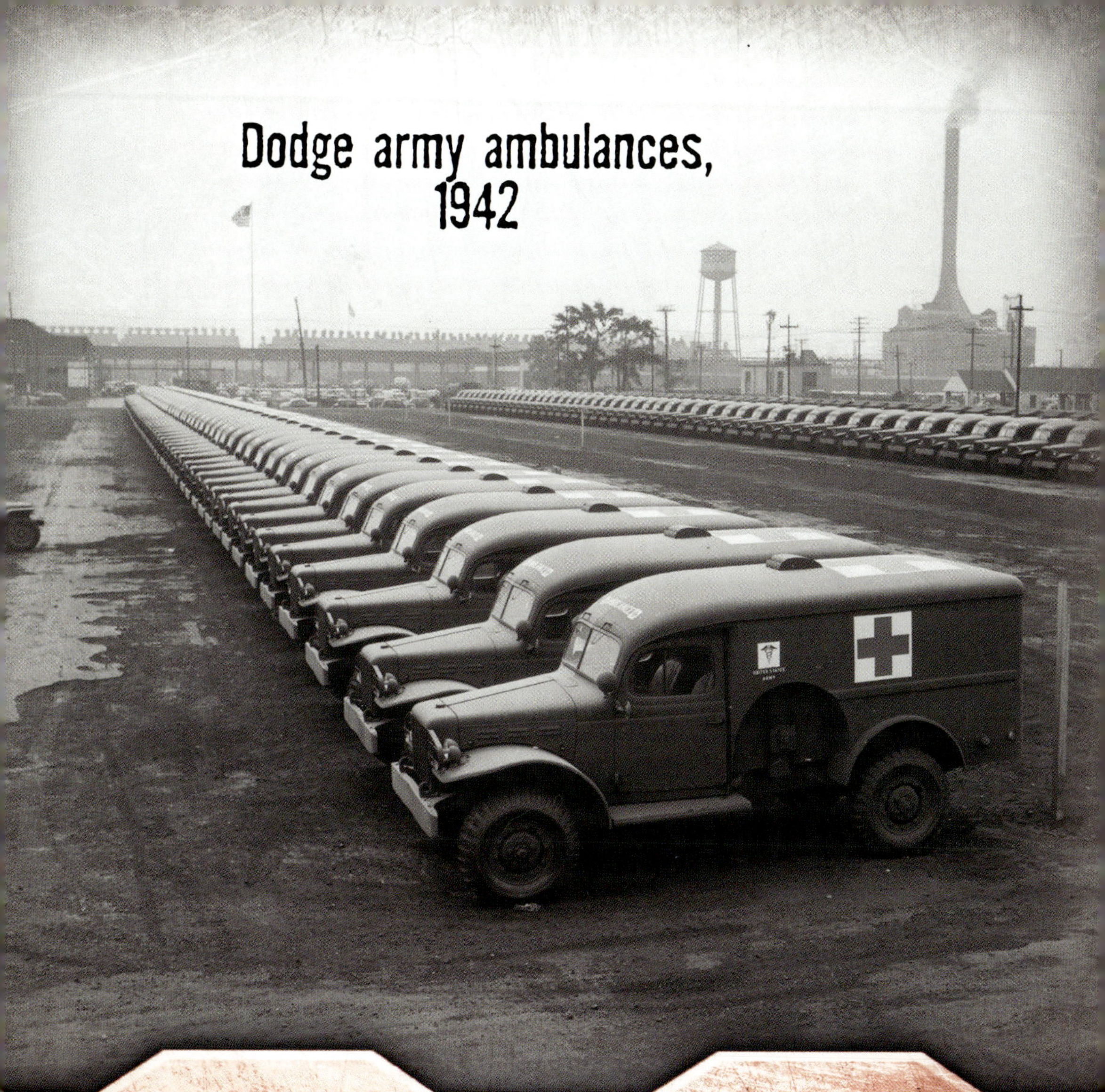

Dodge army ambulances, 1942

FULL THROTTLE!

Dodge used a ram's head on their vehicles from the 1930s to about 1950. It was used again starting in 1981.

Dodge introduced the C-Series pickup trucks in 1954. The C-Series line of trucks had a new frame, new look, and other upgrades that drivers liked. In 1955, Dodge also put out 1/2-ton, 3/4-ton, and 1-ton trucks in the C-Series.

1956 Dodge C-Series pickup

FULL THROTTLE!

The first Dodge trucks
with a **V8** engine
hit the road in the 1950s.

In 1961, Dodge put out the first in a new line of trucks called the D-Series. This included the light-duty D100, the D200, and the D300. By 1963, they would have the first company-built crew cabs, which made them stand out.

1961 Dodge D100

FULL THROTTLE!

Cabs are the front part of trucks where the driver and passengers sit. Crew cabs have four doors, instead of the usual two, and seat four to six people, including the driver.

THE FIRST RAM

The first trucks using the Ram name came out in 1981. The 1/2-ton truck was called 150. The 3/4-ton truck was called 250, and the 1-ton truck was called 350. The first Ram pickups were available as two-wheel or four-wheel drive vehicles.

Ram hood ornament, 1985

1983 Dodge Ram D150

FULL THROTTLE!

In a vehicle with two-wheel drive, the engine sends power to only the front or back wheels. In four-wheel drive, the engine sends power to all four wheels.

The Dodge Ram was a hit! In 1994, Ram pickups got a new look. The different-sized trucks began being numbered as 1500, 2500, and 3500. Dodge Rams were now the biggest trucks on the road!

1994 Dodge Ram D3500 Laramie

FULL THROTTLE!

In 1993, Dodge gave the
Ram a better engine.
With the new look for 1994,
the truck continued to sell well.

THE DAKOTA

While the Ram is the most well-known truck from Dodge, a mid-size truck called the Dakota was on the road starting in 1987. The first Dakotas could come with a 125-**horsepower** V6 engine and four-wheel drive.

Dodge Dakota

FULL THROTTLE!

The Dakota was made until 2011.

THE SPLIT

The Ram trucks were such a success that Ram became its own brand separate from Dodge in 2009. The Chrysler company wanted Ram to only make trucks, sport utility vehicles (SUVs), and other vehicles used for work.

2009 Ram pickup truck

FULL THROTTLE!

Once separate from Ram, Dodge stopped using the ram's head on its vehicles.

RAM TODAY

The 2019 Ram 1500 is built with special steel that makes the frame lighter and stronger. The company calls it the strongest 1500 ever! It's able to carry up to 2,300 pounds (1,043 kg) and tow up to 12,750 pounds (5,783 kg)!

2019 Ram 1500

FULL THROTTLE!

The 2019 Ram 1500 comes in six trim levels, including the Limited, Laramie Longhorn, and Rebel. Trim levels are different features, parts, or abilities available for a type of vehicle.

Depending on which trim level you buy, the Ram 2500 can come with a V8 engine or a diesel engine. It needs its impressive engine—many of the Ram 2500s can seat up to six people!

2017 Dodge Ram 2500
Power Wagon

FULL THROTTLE!

Diesel engines run on diesel instead of gas. Diesel is another type of fuel.

With a 410-horsepower V8 engine, the 2018 Ram 3500 is a powerful truck. It can tow more than 30,000 pounds (13,608 kg)! Ram tests these trucks in heat and on tough hills to be sure they perform well.

2018 Ram 3500 Laramie Longhorn

FULL THROTTLE!

The Ram special edition trucks include the Limited Tungsten, Kentucky Derby Edition, Southfork, and more. They have special looks, colors, and features!

CHASSIS CAB

The Ram Chassis Cab looks like half of a pickup truck. It doesn't have a bed with sides on the back. This line is **configured** to be used for many jobs. Ram Chassis Cabs can become tow trucks, dump trucks, and much more!

2011 Ram 5500 Chassis Cab

FULL THROTTLE!

Just like a pickup truck, the Chassis Cab can come in trim levels that have different features. Drivers choose the trim level based on what job they plan to do.

RAM TRUCKS THROUGH HISTORY

1920s

1920s Dodge truck

1950s

Dodge C-Series truck

1960s

Dodge D-Series truck

1980s

Ram 150

1990s

Ram 3500

2000s

Dodge Dakota

2019

Ram 1500

FOR MORE INFORMATION

BOOKS

Farndon, John. *Megafast Trucks*. Minneapolis, MN: Lerner Publishing Group, 2016.

Mack, Larry. *Ram 1500*. Minneapolis, MN: Bellwether Media, Inc., 2019.

WEBSITES

An Illustrated History of the Pickup Truck
www.caranddriver.com/flipbook/an-illustrated-history-of-the-pickup-truck#1
Find out more about how today's pickup trucks came to be!

Ram Trucks
www.ramtrucks.com
Check out Ram's current truck lineup on the company's official website!

Publisher's note to educators and parents: Our editors have carefully reviewed these websites to ensure that they are suitable for students. Many websites change frequently, however, and we cannot guarantee that a site's future contents will continue to meet our high standards of quality and educational value. Be advised that students should be closely supervised whenever they access the internet.

GLOSSARY

brand: a group of products that are made by a certain company and have a shared name

commercial: having to do with the buying and selling of goods and services

configure: to set something up in a way so that it can be used

horsepower: a unit used to measure the power of an engine

pickup truck: a small truck that has an open back with low sides

vehicle: an object used for carrying or transporting people or goods, such as a car, truck, or airplane

V8: an engine with two banks of four cylinders arranged in a V shape. V6 engines have six cylinders.

INDEX